ISBN 978-1-334-75737-2
PIBN 10573591

For support please visit www.forgottenbooks.com

1 MONTH OF
FREE
READING

at

www.ForgottenBooks.com

By purchasing this book you are eligible for one month membership to ForgottenBooks.com, giving you unlimited access to our entire collection of over 1,000,000 titles via our web site and mobile apps.

To claim your free month visit:

www.forgottenbooks.com/free573591

English
Français
Deutsche
Italiano
Español
Português

www.forgottenbooks.com

Mythology Photography **Fiction**
Fishing Christianity **Art** Cooking
Essays Buddhism Freemasonry
Medicine **Biology** Music **Ancient
Egypt** Evolution Carpentry Physics
Dance Geology **Mathematics** Fitness
Shakespeare **Folklore** Yoga Marketing
Confidence Immortality Biographies
Poetry **Psychology** Witchcraft
Electronics Chemistry History **Law**
Accounting **Philosophy** Anthropology
Alchemy Drama Quantum Mechanics
Atheism Sexual Health **Ancient History**
Entrepreneurship Languages Sport
Paleontology Needlework Islam
Metaphysics Investment Archaeology
Parenting Statistics Criminology
Motivational

United States Department of the Interior
Fish and Wildlife Service

Fishery Leaflet 439 — — —

THE DUNGENESS CRAB INDUSTRY

By Fred W. Hipkins[1]

The Dungeness crab (Cancer magister), often called the Pacific crab, is a large hard-shell crab measuring up to 10 inches across the carapace (fig. 1). It is named after a small fishing village, on the Strait of Juan de Fuca in Washington, where the commercial fishing for this crab began. The Dungeness crab, in its several market forms, is considered a choice fishery product of the North Pacific Coast and provides a highly palatable protein-rich food.

Figure 1 - Dungeness Crab

The Dungeness-crab fishery extends from San Francisco Bay in California to Prince William Sound in Central Alaska and provides seasonal employment for more than 5 thousand persons working in about 95 shore plants. Nearly a thousand fishing boats and vessels land about 28 million pounds of Dungeness crabs annually. The combined assets of vessels, fishing gear, and shore plants are valued at more than 10 million dollars.

----

[1] Fishery Marketing Specialist, Bureau of Commercial Fisheries, Fishery Products Laboratory, Ketchikan, Alaska

Although the Dungeness crab is found along thousands of miles of coastal areas it is only known to inhabit sandy and grassy bottoms below the tidal range. Such bottoms are widely scattered along the coasts of Northern California, Oregon, and Washington. The principal landing ports are San Francisco and Eureka in California, Coos Bay, Yaquina Bay, and Astoria in Oregon, Grays Harbor, Willipa Harbor, and Seattle in Washington, and Cordova, Petersburg, Kake, Hoonah, and Ketchikan in Alaska.

In recent years Dungeness-crab fishing has become an important industry along the coasts of British Columbia, Canada. The northern part of Vancouver Island, Hecate Strait, and Dixon Entrance provide very productive fishing grounds. The latter two areas are also fished by American vessels, which deliver their catches to Seattle, Wash., or Ketchikan, Alaska.

Table 1 shows the landings of Dungeness crabs in California, Oregon, Washington, and Alaska from 1946 to 1954.

Table 1. -- Landings of Dungeness crab, 1946 to 1954
(In pounds)

Year	California	Oregon	Washington	Alaska	Total
1946	9, 644, 400	7, 742, 400	8, 008, 400	2, 438, 600	27, 833, 800
1947	10, 733, 400	7, 531, 700	11, 973, 200	1, 392, 611	31, 630, 911
1948	11, 818, 300	10, 069, 500	22, 712, 500	1, 222, 326	45, 822, 626
1949	11, 116, 300	9, 346, 500	13, 142, 500	1, 428, 401	35, 033, 701
1950	11, 722, 500	6, 266, 800	5, 821, 100	4, 119, 425	27, 929, 825
1951	11, 568, 400	7, 452, 000	4, 829, 968	5, 482, 416	29, 332, 784
1952	12, 997, 400	5, 456, 700	4, 265, 800	3, 749, 412	26, 469, 312
1953	8, 277, 714	7, 824, 000	7, 004, 090	3, 471, 806	26, 577, 610
1954	7, 828, 000	9, 602, 000	8, 311, 000	2, 739, 383	28, 480, 383

2

HISTORY OF COMMERCIAL FISHERY

The Dungeness-crab fishery is both the largest and the oldest known shellfish fishery of the North Pacific Coast. Many years before the arrival of the white man, natives along the shore of the Strait of Juan de Fuca trapped crabs in crude homemade pots. The settlers were quick to realize the value of the Dungeness crab and began fishing for them.

In California, the Dungeness-crab fishery began between 1860 and 1870 in the San Francisco Bay and Golden Gate areas. No restrictions on the taking of crabs were then considered necessary. Later, the fishing so increased that the California Legislature in 1903 prohibited the taking or possessing of female crabs, and made it illegal to take any crab less than 6 inches across the carapace. In 1911 the size limit was amended to 7 inches, and that limit is still in effect.

CRAB FISHING

Method of capture

"Pots" and ring nets are the only types of fishing gear authorized for the commercial taking of Dungeness crabs in the Pacific Coast States and Alaska. Pots are individual traps fished at fixed positions on the bottom. The construction and operation of crab pots is explained in Fishery Leaflet 419, "Dungeness Crab Pots", which can be obtained free of charge from the U. S. Fish and Wildlife Service, Department of the Interior, Washington 25, D. C.

Fishing areas

The Dungeness-crab fishery is carried on both in inside waters like San Francisco Bay, Puget Sound, and Prince William Sound, and in outside waters.

Fishing depths generally range from 2 to 20 fathoms, but occasionally crabs may be fished in depths up to 40 fathoms. During the fishing season, fishermen move their pots according to the movements of the crabs: one week, crabs may be found in depths of 2 fathoms; the following week, they may be found in depths of 15 or 20 fathoms. The crabs' location is generally determined by the trial setting of ring nets in different depths of water.

Fishing Boats

        Crab boats or vessels vary from 16-foot outboard-powered skiffs to 65-foot seine-type or otter-trawl vessels. The smaller boats normally confine their operations to the inside waters, although any type of boat can be used there. Boats in the inside fishery do not have far to travel and can deliver their catches within a few hours. In recent years, vessels of the larger type have been traveling as far as 600 miles to the fishing grounds (fig. 2). These boats have built-in holding tanks supplied with circulating sea-water in which the captured crabs can remain alive indefinitely.

Figure 2 - Crab pots loaded aboard a barge for transport from Washington to Southeastern Alaska waters.

4

Regulations

Regulations for the commercial taking of Dungeness crabs are set by the regulatory agency having jurisdiction over the area where fishing or landings occur. The fishing seasons open and close in different months of the year according to the area, with the exception of the Columbia River (Astoria) area, which is open the entire year. There are three regulatory measures common to the three Pacific Coast States and Alaska: (1) The taking of female crabs is prohibited; (2) the taking of crabs with soft shells is prohibited; and (3) the commercial taking of Dungeness crabs in fishing gear other than pots and ring nets is prohibited.

Regulations governing the size of crab vary. In Alaska and California, the minimum size is 7 inches across the top of the carapace, between the extreme points of the lateral spines. Oregon regulations specify $6\frac{1}{2}$ inches, and Washington regulations $6\frac{1}{4}$ inches minimum size; both Washington and Oregon require the measurement to be taken across the back or top of the carapace and immediately in front of the lateral spines.

PROCESSING

Crabs are delivered alive to shore plants, generally located on docks convenient to unloading facilities (fig. 3). Crabs must be kept alive until they are processed. Large crabs with the claws and legs intact are generally processed whole for the fresh-market trade. Crabs without legs or claws are processed for picking, and the meat is marketed fresh, frozen, or canned.

Figure 3 - Live crabs in holding tank ready
for unloading at shore plant.

5

Male crabs, when trapped in pots often lose legs and claws from fighting. Care must be used in the handling of the crabs to avoid injury, because they are especially quick with their claws.

Depending on the climate, crabs will remain alive from 2 to 8 hours after being removed from the water. Dead crabs are not acceptable at any processing plant. Upon arrival at the plant, if the processing does not begin at once, the crabs are placed in holding tanks (fig. 4).

Figure 4 - Floating holding tanks moored near the
shore plant.

Whole Crabs

Crabs selected for the fresh-market trade are boiled in a brine solution (3 percent salt by weight) from 15 to 20 minutes, depending on the size of the crabs. After cooking, they are chilled in fresh cold water, and the shell is brushed to remove any particles that may cling to the outer shell, thus giving a more attractive appearance. The whole crabs are then ready for the market.

Crab Meat

Crabs selected for picking are butchered by removing the carapace or back shell, the viscera, and the gills (fig. 5).

Figure 5

Plant scene, showing the three stages in butchering of crabs.

If the crabs are thoroughly bled and cleaned, there is less chance that blood residues will cause undesirable color changes in the flesh during the cooking. The crabs are next broken into two sections which are carried on a conveyor belt through a cooking chamber containing water at 210° to 212° F., or steam at 212° to 220° F. (fig. 6).

Figure 6 - A steam-heated cooking chamber.

The cooking time is 10 to 12 minutes. The cooked sections are sprayed with clean fresh cold water immediately upon leaving the cooker. This cools the crab sufficiently for the shakers to handle them, and tends to loosen the meat from the shell.

In some plants a batch-cooking process is used. The crabs, in baskets, are lowered into a vat of boiling water where they are cooked for 10 to 12 minutes. The baskets are then lowered into a second vat of clean cold water where the crabs are momentarily cooled.

The still-warm sections of the cooked crabs are taken to the shaking tables where the meat is removed from the shell. Women employed to remove the meat are called "shakers". The meat from the body cavities is removed before the leg meat. To remove the body meat the shaker lightly crushes the body between the palm of her hand and the table to loosen the meat imbedded in the cavities. The crab section is then tapped against the side of the pan and the meat falls into the pan (fig. 7).

Figure 7

Meat from the body sections are shaken out by workers.

The leg shells are then cracked with a small metal mallet. The legs are torn from the body shell, broken off at the joints, and each section is tapped against the side of a pan to shake out the section in a whole piece. The leg meat and body meat are shaken into separate pans. An experienced worker can shake from 150 to 200 pounds of meat during one 8-hour shift and normally is paid at the rate of 10 cents a pound (the rate of pay varies between locations).

Time-saving improvements in crab meat processing have been made in the past several years. One of the major improvements is the use of the brine-flotation method for washing the meat. This operation serves a threefold purpose: (1) it removes most of the shell fragments and tendons that may be mixed with the meat; (2) it washes the meat; and (3) it flavors the meat with salt. During this process the brine solution is constantly agitated, so that the meat rises to the top and the shell fragments and tendons sink to the bottom. Although this treatment is a big improvement in processing techniques and saves time in sorting out shell and tendons, it does not completely eliminate all of the shell fragments and tendons. The meat is therefore thoroughly inspected as it comes from the brine tank on a small conveyor. The meat next receives a short fresh-water rinse as it goes up the conveyor into draining pans where it stands until all excess moisture has been drained off (fig. 8).

Figure 8

The meat is given a brine-flotation treatment, inspection, and rinse.

9

An acid dip is generally used when processing plants employ the batch-cook method for cooking crab sections. The outside of the body-cavity meat that is exposed to the boiling brine, often turns a slight-gray color when cooked by this method. To minimize off-color, the meat is generally dipped for about 1 minute into a dilute organic acid solution which acts as a bleach. Citric acid is preferred, although acetic acid or lactic acid is also used.

After the excess moisture from the rinse or the acid dip has been drained away, the meat is taken to the packing room and packed in containers for either the fresh-and-frozen trade or for canning. For the fresh-and-frozen trade the meat is normally packed in No. 10 C-enamel (seafood formula) cans and sealed with a tight-fitting cover. The bottom half of the No. 10 can is filled with body meat and the top half with leg meat. The leg meat is considered a higher grade than the body meat, since it has a better appearance, with streaks of orange and purple color, and generally has a firmer texture.

All of the equipment pans -- mallets, table tops, and other surfaces -- that comes in contact with the meat is made of stainless steel. At the close of each day's operation, all of the equipment is thoroughly washed and sterilized.

Canning

Since 1952 most canned crab meat has been packed in the standard tuna-size can, containing about 6 ounces of meat (fig. 9).

Figure 9

Crab meat hand-packed in containers for canning.

Some packers still prefer to use the half-pound flat salmon cans, and some crab meat is also canned in the 1-pound flat size. Regardless of size, all cans are treated inside with the C enamel. The cans are normally packed with a layer of leg meat on both bottom and top, the center being filled with body meat. Some packers may pack the can with leg meat in only the top layer. After the cans are filled, they are hermetically sealed under vacuum, suitably heat-processed, then chilled quickly with cold water.

More information about the processing and canning of crab meat is found in Fishery Leaflets 85, 88, and 374, and Separate No. 50, available free of charge from the U. S. Fish and Wildlife Service, Department of the Interior, Washington 25, D. C.

MARKETING

Dungeness crab is marketed in three different forms: (1) Whole crab, fresh or frozen, (2) fresh or frozen meat, and (3) canned meat. Nearly 85 percent of the total catch is marketed fresh and frozen, either as whole crab or as picked meat, in States west of the Rocky Mountains, particularly in the Pacific Coast States. Nearly all the crabs landed in California, the largest marketing area on the Pacific Coast, go to the fresh-market trade. Fresh-crab shippers in Oregon and Washington also supply the California market. A few individually packaged and frozen whole crabs are marketed in the Midwestern States as well as in some of the large eastern cities. Canned Dungeness crab meat is offered for sale throughout the United States.

Before 1920, Dungeness crab was marketed in one form only -- freshly cooked whole crab. Up to that time, the domestic canned crab meat market was completely dominated by foreign crab meat, mainly from Japan. Domestic producers, lacking a good formula, were skeptical about canning the meat and felt that they could not compete with foreign producers. For several years before 1920, trade journals strongly emphasized that Dungeness crab meat had a flavor superior to foreign canned crab meat and had urged the crab industry to start canning Dungeness crab. It is claimed that a canning formula was borrowed from Japan, and with the encouragement of a trade journal, Dungeness crab meat was first canned in Alaska in 1920, and in Washington and Oregon in 1927.

The small Alaska pack of 75 cases in 1920 was a success, and the industry in Alaska quickly expanded. Canners were quick to enter this new field because it afforded off-season employment to cannery workers and fishermen. In 1948 a record pack of 169,798 cases of Dungeness meat, valued at $3,820,622, was produced by the entire industry. This amount represented about 30 percent of the crabs landed that year, while the remaining 70 percent was marketed fresh or frozen as either whole crabs or picked meat.

11

In recent years, the production of canned crab meat has declined. Two factors have caused a reduction in the production of heat-processed canned crab meat: (1) The demand for fresh and frozen crab has steadily increased, and (2) increased production and importation of canned crab meat, mainly from king crabs, have caused the domestic industry to proceed cautiously. Table 2 shows the canned domestic Dungeness crab meat pack from 1948 through 1955.

Table 2 - Annual pack of canned Dungeness crabs in standard cases of 48 half-pound flat cans - 1948 to 1955

Year	Alaska	Washington	Oregon and California*	Total
1948	8,454	104,362	56,982	169,798
1949	13,198	65,004	36,652	114,854
1950	25,636	23,270	29,335	78,241
1951	26,697	19,885	42,715	89,297
1952	16,276	9,076	9,861	35,213
1953	14,547	15,482	8,044	38,073
1954	8,425	36,141	25,079	69,645
1955	6,548	9,450	3,115	19,113

* Since the California pack was canned by a single firm, the production has been included with that for Oregon.

WS - #0118 - 200624 - C0 - 229/152/1 - PB - 9781334757372 - Gloss Lamination